16
A WRIGLEY BOOK
about
TIME

BY DENIS WRIGLEY

Will it take long
to walk to the end of the road?

NOT LONG!

Will it take long
to drive to the town?

NOT LONG!

Long distance — fast car. . . .
short distance — slow legs. . . .

What does 'not long' mean then?

Everything takes time to happen. . . .

A game takes time to play.

It takes time to go to school
and it takes time to go on holiday,

and we can measure time
by dividing it into parts.

We call them seconds and minutes,
hours and days and weeks
and months and years.

We mark these measurements on
clocks and calendars
and in diaries.

Now, things that have already
happened, like the game
you played yesterday,
can't happen again.
We say they happened in the PAST.

The game you are playing now
is happening in the PRESENT
and the game you may play tomorrow
will be in the FUTURE.

So PAST, PRESENT and FUTURE
describe when, but not how long.

We can now be exact
and say things like,
'See you in five minutes *exactly*'
And in five minutes *exactly*
I'll be there. . . .

if I'm not late!

Being late means you haven't been
exact in your timing;
being late means
you took too long,
were too slow in arriving.

Calling things slow or fast is not
a measurement of time,
it is only a way
of comparing the speed
of one thing with another.

A car is faster than I am,
it travels the same distance
in a shorter time.

I am slower than the car.
It takes me a longer time
to walk the same distance.

So when we say 'NOT LONG'
we mean not a long time
compared with the time
it would take something else
to happen.

It will not be long till
dinner time. . . .

(Compared with the time
till tomorrow's dinner time)

It will not take long
to walk to the end of the road.

(Compared with crawling!)

It will not take long
to drive to town.

(Compared with *driving*
a hundred miles
or *walking* to town)

You can waste time,
you can lose time,
you can save time.

Think some more about time. . . .

It's time you did!

First published 1976
Copyright © 1976 Denis Wrigley
ISBN 0 7188 2188 2
Printed in Hong Kong

The Wrigley Books

Published by
LUTTERWORTH PRESS • GUILDFORD AND LONDON